How to Launch a New Product Faster

5 Secrets to Tight Project Management, Comprehensive Project Plans, and Effective Change Management

Elizabeth Cavanagh

How to Launch a New Product Faster: 5 Secrets to Tight Project Management, Comprehensive Project Plans, and Effective Change Management

ISBN: 978-0-615-60459-6

Cover and interior artwork by Daniel Tero

Published by Rockland Publishing, Danville, CA

Printed in the United States of America

DEDICATION

I dedicate this book to my family

because they make my heart smile everyday

ACKNOWLEDGEMENTS

Let me start by acknowledging the impact my brother, John, had on me writing this book. He is an amazing entrepreneur with a passion for helping small businesses hone their marketing efforts. And with this knowledge, he reminded me to view this book not from my perspective, but from my reader's perspective. He was always available to talk and kick around ideas, and he always provided such insightful recommendations. So John, thank you.

My Mom and Dad taught me a lot about the business world and reinforced that we should deal with office politics directly and with diplomacy. Many Saturday mornings were spent drinking coffee, discussing how to handle different political situations at work. So for the advice and the foundation, thank you Mom and Dad.

My friends, Wendy and Barb, helped edit this book with razor-sharp eyes and amazing objectivity, and allowed me to push this book to the next level. They continually encouraged me and believed in me. So, to Wendy and Barb, thank you.

CONTENTS

How to Launch a New Product Faster

*5 Secrets to Tight Project Management,
Comprehensive Project Plans, and
Effective Change Management*

Introduction

Introduction

In this book, I'll walk you through a collection of *Secrets* that I have uncovered after leading or participating in a number of high-tech product and marketing launches over the last 20 years.

I call them Secrets because I've applied them to product launches time and time again only to get the same result – I was able to *guide* large teams to successfully launch new products very quickly.

I have organized these Secrets into a chronological 5-step blueprint to allow others to easily learn about them, and understand how and when to properly apply them.

Why use these Secrets? They will teach you how to avoid   common   mistakes,   how   to   increase   team

productivity, how to use senior executive involvement to your advantage, and as a result, how to launch new products faster.

In the book, *Execution: The Discipline of Getting Things Done*, by Larry Bossidy and Ram Charan, they state, "Everybody talks about change… But unless you translate big thoughts into concrete steps for action, they're pointless. Without execution, the breakthrough thinking breaks down, learning adds no value, people don't meet their stretch goals, and the revolution stops dead in its tracks. What you get is change for the worse, because failure drains the energy from your organization. Repeated failure destroys it."

By continuing to launch new products without concrete steps, you run the risk of draining yourself, your project team and your executives, and you stop adding value.

In this book, *How to Launch a New Product Faster,* you will learn the 5 steps to effectively project manage product launches. The benefits don't stop there because this is a standard, repeatable process you can use in *all* product launches or other marketing-related launches. I've even used it to help me lead more general change management efforts.

What makes this book different from the other project management books? It covers both the *Process* and the *Politics* of change management. What this means is that this book gives you a standard 5-step process to follow to successfully manage all the work that needs to get done, and it also provides a structure to help you manage the political roadblocks that every project encounters. By managing both of these elements, you are well on your way to a tightly managed and successful product launch.

> ### *Key Point*
>
> *I have added 'Key Points' throughout this book to highlight the key lessons that are critical to deploying each of these Secrets. They will be easy to identify – just look for the shaded box and the graphic of the person holding a checkmark.*

Below are three different types of Project Managers who specialize in launching new products. Which type are you?

- Those who know how to launch products and want to learn how to *standardize* the process and manage the politics of implementing a change -- They will benefit from this book by adding a standard approach to what they already know.

- Those who have launched products in the past and happened to get it right, but want to learn a standardized approach based upon Best Practices so they can get it right every time -- These people will benefit from this book by learning a standard, 5-step, simple process to become more efficient and give them more confidence at being a change leader.

- Those that don't know how to execute a change within a company and have failed at launching products in the past. They are hesitant to volunteer to lead new projects. They are aggressively seeking out information on this topic -- They will benefit from this book by learning how others have successfully become change leaders by learning the 5 steps to successfully launch a new product.

No matter which group you fall into, this book offers a new perspective on how to hone your project management and change management skills. It offers concrete examples, walks you through a standardized, best practice, 5-step process for launching a new product (and guiding your product launch team), highlights Key Points to add further clarity, and gives insight on how to handle the politics along the way.

As you read this book, please keep in mind that the secrets and methodology described are actually a **Blueprint for Managing Any Large Change**, *not just new product launches.*

These 5 steps can be applied in the business world to help **guide** a business change as well as in your personal life to help guide a change you want to make in your life.

Now, let's get started and get you on your way to learning *How to Launch a New Product Faster.*

The Value of Using These Secrets

Chapter I

Why Use These Secrets

Why is it that most ideas are never implemented? Even when they could change the direction of a company? It's because these ideas are not managed. And if it's not managed, it won't happen.

And the reverse is true - What's properly managed is what actually happens.

What drove me to write this book is that I've been in too many meetings where people talked about wonderful new product ideas -- some which would save money, some which would generate additional revenue, and some which would greatly improve customer satisfaction. But most of these ideas never went anywhere because they were not *managed*. **What's properly managed is what actually happens.** This belief is what forms the basis of these 5 Secrets.

> **"We are continually faced by great opportunities brilliantly disguised as insoluble problems."**
>
> Lee Iacocca

Over my years of working on product launches, I have found that there are common mistakes that people make which derail their efforts. At a *strategic* level, the most common mistakes with product launches center around the lack of strategic communication. At a *tactical* level, the most common mistakes center around the lack of tight project management.

The 5 Secrets were created to help others avoid these common product launch mistakes:

- There is not a well *documented* and *agreed to* goal
- All the *changes* that are required to launch the product are not properly *identified*, including process changes, technological changes and resource changes
- They do not have *tight project management* of these changes, including tight launch team and project management
- They do not have a formal, data-based *go/no-go launch decision* point with senior executives
- They do not *measure the success of the launch (pre vs. post launch results)*
- They do not provide effective, strategic communication about the purpose of the product launch and what is required from senior management to ensure success
- They do not manage the politics that come with any change effort directly and with honesty

Many people believe that after it is decided which product to launch, the work is done. However, there is a lot to do to create the product and get it to market. After reading this book, you should have a clear understanding of the most common mistakes project managers make when launching new products, and how these 5 Secrets will help you avoid these mistakes and enable you to launch new products faster by more tightly **guiding** your launch team. It's so easy to get derailed, but by using these 5 Secrets, and understanding how they work to manage both the work and the politics, you can more easily guide your project team to a very fast and more successful launch.

The 5 Secrets

Chapter II

5 Secrets to Launch a

New Product Faster

Think about the people in senior positions in your company. They come from varied backgrounds with varied expertise being developed over the years. However, they all have one thing in common – they know how to execute change. They know how to take the company's business plans and put together supporting programs that move the company forward toward their goals. In short, they know how to create and launch new ideas, and because of this, people want to work on their teams.

Leading a business change is hard work. Not many people like change. As a change leader, it is your job to convince them that this change is a good idea, that their role in making the change happen is critical, and that they need to help others see this as a positive thing and to embrace the change.

Launching new products is just as hard, and I have uncovered the 5 Secrets to Launch a New Product Faster, which, in essence, is my 5-step launch methodology. By following these 5 steps, you will be able to take on enormous projects and smaller ones without having to recreate the process to execute the change. You will also be given guidance for managing the politics along the way. As stated earlier, not many people like change and, therefore, you'll have to manage the politics of getting people on board early in the project. Not to worry, these techniques have been proven and are simple to follow. Just remember that everyone *wants* the company to succeed, so at the end of the day, you all have the same goal in mind – your company's success.

I'll now describe the 5 Secrets to Launch a New Product Faster, and in the following chapters, I'll go through each Secret, or guiding step, in detail. I will also discuss, with each step, how to manage the politics.

As you read through this book, keep in mind that these are guidelines and that each project is different. Some projects are quite large in scope and need to follow a rigorous change management approach as outlined in the book. Other projects are small and can be managed by scaling down this approach.

5 Secrets to Launch

a New Product Faster:

There are 5 Secrets to simply and quickly *guide* the launch of a new product:

1. **G***round* Your Project with A Clear Goal

2. **U***nderstand* the Required Changes

3. **I***ntroduce* Tight Project Management

4. **D***rive* a Data-Based Launch Decision

5. **E***stablish* Pre- and Post-Launch Measurements

In order to make these 5 Secrets easier to remember, I've created an acronym – the word 'GUIDE' – since it spells out the first words of each Secret. I chose this word because that is exactly what a launch project manager does, guide his or her team to launch a new product or program.

More detail will follow describing each of these 5 Secrets later in the book. However, I want to emphasize the importance of one Secret in particular. The third Secret, 'Introduce Tight Project Management,' is the largest and most time consuming step in your launch process and needs to be tightly managed. It has been my experience that this is where most executions fail. Why? Because it is *the most tedious step to manage*. This chapter highlights how to manage all the moving pieces of work while managing the people doing the work, and how to handle the politics that sometimes swirl around the people doing the work.

So, be patient with yourself when you come to using tight project management and it will definitely pay off. Remember, **what gets managed gets done**. To help you with this, pay close attention to the best practice techniques and tools in Chapter 5.

How the Secrets are Revealed

To better explain these 5 Secrets (or **guiding** steps), we'll use an example of a fictitious software business and walk you through how a project manager uses these 5 Secrets to successfully *guide* the launch of a new software product.

The *fictitious* company we will use is called Software For Business Inc., and they are launching a new software product called 'The Tracker,' which will help other companies better track their expenses. The perspective that will be used in this book is that you are the project manager for 'The Tracker' and are guiding your launch team through each of these 5 steps to launch this new product quickly and get incremental revenue for your company sooner.

5 Secrets to *Guide* a

Faster Product Launch:

1. **G***round* Your Project with A Clear Goal

2. **U***nderstand* the Required Changes

3. **I***ntroduce* Tight Project Management

4. **D***rive* a Data-Based Launch Decision

5. **E***stablish* Pre- and Post-Launch Measurements

Chapter III

The First Secret:

Ground Your Project With A Clear Goal

Don't underestimate the importance of a clear goal. A clear goal sets the pace, tone and direction for the entire team, and helps to manage expectations of senior management.

If your goal is too broad, too many different interpretations can happen leading to disparate team directions. If your goal is too narrow, creativity may be blocked.

Once it is defined, make sure the entire team has a clear understanding of the overall team's goal. They'll use this knowledge to make decisions and tradeoffs within their own functional areas. By understanding the team's goal, your individual team members can make more informed decisions and trade offs that are aligned with the overall team goal.

In her book, *Real People, Real Change*, Donna Strother Highfill says, "As change warriors, it is our job to help employees identify and agree upon one common goal, and then create a new road to get there." This is the basis of change management.

To begin drafting your goal, meet with the champion of the new product (the project sponsor). This is the person who is the overall champion of the new product within your company, the person who has the most to loose politically if the product doesn't get launched.

To **clearly articulate your goal**, work with your project sponsor and make sure your goal is specific and includes the following:

- Rationale (why it's important)
- What will be delivered (product view)
- With what financial results, short- and long-term views (revenue results)

- Within what time period, short- and long-term (by when it needs to be accomplished)
- Based on what assumptions (parameters)

Make sure your goal is achievable by checking it against the project's:

- Staffing Availability
- Funding
- Technology Capabilities
- Overall Timing
- Strategic 'Fit' within the Company

Once finished, the next step is to properly socialize and solidify your goal with those affected and those against it (more on this to follow later in this chapter).

Let's use our example company, Software for Business Inc., and walk through the steps to clearly articulate our goal to launch The Tracker. I'll provide both a good example of a goal and a bad example to highlight the differences.

Here is a Good Example of How to Articulate a Goal:

Software for Business Inc. will launch a new software package called 'The Tracker' on October 15, 2012 (six months from now) in order to capture the untapped market of small and medium businesses that are not able to easily track their expenses without purchasing very expensive, large-scale software programs. By launching this product, we will capture $1M in incremental revenue within a year, and $5M within three years.

Key assumptions include: team will get proper staffing and funding; IT will be able to meet software development changes on schedule; vendors will be able to supply needed components; the product will be distributed through existing distribution channels; product will be priced at a mid-market price; product will be marketed as all new products are marketed, though targeted only to small and medium businesses.

Here is another way to check that your goal includes all the right components:

Goal Should Include:	Example:
Rationale (why it's important)	Capture the untapped market of small and medium businesses that are not able to easily track their expenses without purchasing very expensive, large-scale software programs.
What will be delivered (product view)	Software for Business Inc. will launch a new software package called 'The Tracker' on October 15, 2012 (six months from now)
With what results – short- and long-term (revenue view)	We will capture $1M in incremental revenue within a year, and $5M within three years.
Within what time period –short- and long-term	See above
Based on what assumptions (parameters)	Team will get proper staffing & funding; IT will be able to meet software development changes on schedule; vendors will be able to supply needed components; the product will be distributed through existing distribution channels; product will be priced at a mid-market price; product will be marketed as all new products are marketed, though targeted only to small and medium businesses.

This is a good example because it explains what the team is focused on from a product view as well as from a short- and long-term revenue view, why it is important, and when it needs to be accomplished. This will allow team members to have a consistent vision of where they are headed and why they are headed there.

By including longer-term revenue goals, the project team members can lay the infrastructure and foundation to help make these longer-term goals happen, and there is a vision on which to base longer-term decisions. By including assumptions, you communicate the parameters under which your goal is being executed to better manage expectations and minimize political discourse further down the road.

Here is a Bad Example of How to Articulate a Goal:

Launch a new software product to bring in more revenue without adding new distribution channels.

This is a bad example because the team members won't have a common understanding of the rationale, what will be delivered, with what results, within what time period, and based on what assumptions. It also does not lie out any future goals so the groundwork elements (infrastructure, resources) won't be properly laid out for future phases of work.

Socialize Your Goal

Once your goal is clearly articulated, socialize your goal with the following three groups of people to gain buy in and support. Members of all three groups can come from inside or outside the company, but ideally they're inside. The three groups are:

- Senior Executives affected by the change; ask your Project Sponsor for names; these will make up the Executive Steering Committee (ESC) – more on this team later in this chapter
- Project Team members (working level team)
- Subject Matter Experts (SME's) affected by the change (ask team members for names)

Ask the Senior Executives to validate that the team goal is appropriate, fits within the overall strategic plan of the company, and fits with the overall priorities of resources within the company. This information will help you better explain it to others.

Also, be sure to continually ask the Senior Executives "why" to better understand why this goal is important to the company. After the Senior Executives validate the team goal, the next step is to have your project team members and others affected by the change validate that this goal is reasonable and attainable once your project team has been formed.

If you don't have buy-in from these three groups, they won't take ownership and you will not have an optimized team. Make sure the *whole* team is truly on board before moving ahead. If you run into a roadblock, ask for help from your project sponsor.

During this process, expect there to be differing opinions that will result in changes and/or tweaks to your goal. Although it might be frustrating and time consuming, it's better to get differences identified and addressed sooner rather than later in order to gain alignment and build a good foundation for change. This is what helps to minimize the politics. Keep in mind that any changes made to the goal must be re-routed through these 3 groups as needed to keep everyone on the same page and aligned.

> ## *Key Point*
>
> *Expect your project to have issues*
>
> *(managing them is the key to effective*
>
> *change management) and work to*
>
> *uncover these issues early in the*
>
> *product launch cycle versus waiting*
>
> *further down the road.*

How Different Teams Work Together

On the next page is a typical Team Governance structure that has been created for our fictitious software product launch, The Tracker.

This type of governance structure has been used again and again, and it is a proven best practice because it provides teams with upper management support for clearing roadblocks while at the same time providing a structure for subject matter experts to provide their expertise *only* when it is needed.

Typical Team Governance Structure

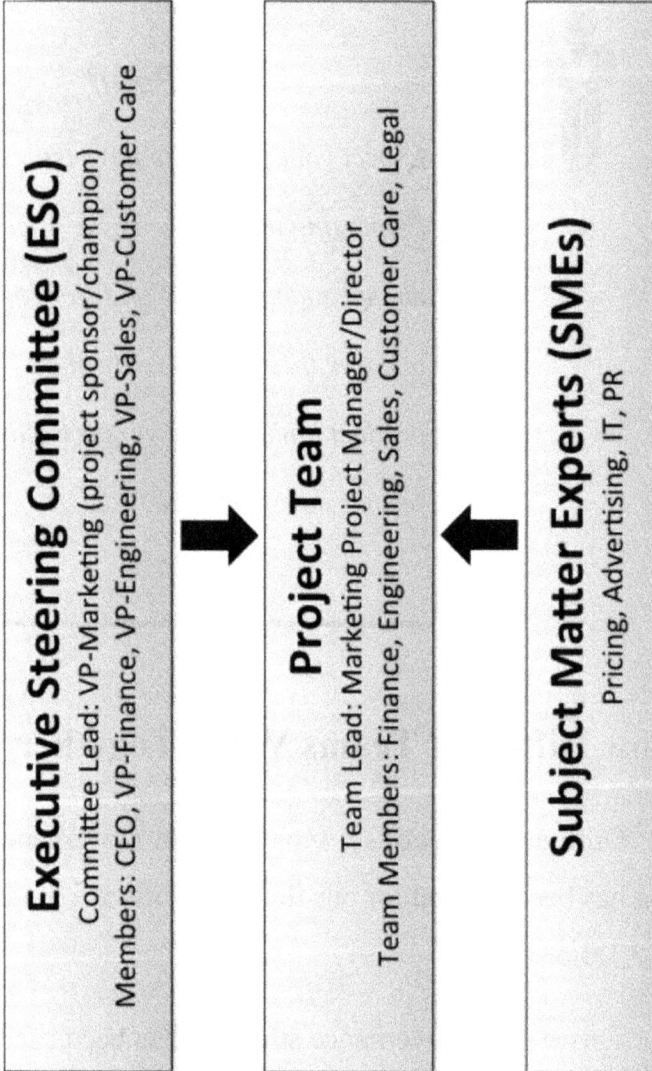

Executive Steering Committee (ESC)

Committee Lead: VP-Marketing (project sponsor/champion)

Members: CEO, VP-Finance, VP-Engineering, VP-Sales, VP-Customer Care

Project Team

Team Lead: Marketing Project Manager/Director

Team Members: Finance, Engineering, Sales, Customer Care, Legal

Subject Matter Experts (SMEs)

Pricing, Advertising, IT, PR

The larger your product launch, the more intricate the separate teams. Whereas if your product launch is simple and straight forward, your governance structure can also be simple and straight forward.

Below we describe the steps to create your Team Governance structure. Start by meeting with your project sponsor and create a list of VPs and key influencers from the functional areas that are most affected by your goal.

Why do you need an Executive Steering Committee (ESC)? They provide you with several advantages. They:

- Provide buy-in of your goal at the senior level
- Help steer hard changes through the company
- Are responsible for managing escalations (issues that need to be handled at a higher level)
- Provide a more strategic view of how this change 'fits' within the overall company plan
- Provide dedicated team members and subject matter experts from their functional areas

In the example of introducing a new software product, the following people would be on the ESC:

- CEO
- VP-Finance

- VP-Engineering
- VP-Sales
- VP-Customer Care
- VP- Marketing (usually the project sponsor)
- Any other senior executive who is a key influencer

It is also helpful to pull in anyone who normally would get in the way of reaching your team goal. This may sound counter-intuitive, but it does two things: 1) it helps win over your strongest opponent sooner rather than later in the product launch process, and 2) it helps make them accountable for helping you attain your goal.

It is all right to involve people outside your company if they will be helping to make or are affected by the change. In our example of the new software product, assume that an outside vendor provides the sales function. A senior level representative from this vendor would sit in for the VP-Sales position on the ESC, and working level vendor employee(s) would be on the project team as the sales team lead.

The ESC should be updated on the team's progress once a month, depending upon the scope of the changes needed. This will help to manage the politics. More details about how to best utilize your ESC is provided in later chapters.

When creating your governance structure, make sure every team member understands their role to ensure that they will *own* their role. While this process may be time consuming, it is invaluable later in the project life. More on this on the next page.

Staff Your Project Team

During the process of socializing your goal with your ESC, ask each member who he or she wants (from within their functional area) to be on your Project Team. They can be from within the company or, if appropriate, an outside vendor. This lays the foundation for forming your working level project team. Also, ask who should be brought in as subject matter experts so you know whom to tap when needed. Staffing your Project Team from the top down vs. approaching them directly sets the tone that this is a strategically important team, and lends more credibility to the purpose of the team and the work of its members. It also helps better manage the politics.

Once your project goal has been socialized with the ESC, and your team members and SMEs have been identified, you can begin the work of formally kicking off your Project Team.

Kick Off Your Project Team

As you begin to form your Project Team, meet with each member and identify *specifically* what role he or she will play and ensure you have his or her buy-in. Start by first painting the overall picture of what the team is trying to do (the project goal) and how they *specifically* can help advance this goal (their role). Then, take a step back and check to see if you have their buy-in. If you don't have it, seek to understand what issues they have and work to get rid of these issues as quickly as possible.

Every week, hold a Project Team Meeting for all key players on your team. (More detail about these meetings and a template for a standard weekly meeting deck is provided under 'Weekly Project Team Meeting' in Chapter 5.) A weekly meeting will help you stay on top of key issues and help team members feel accountable for advancing their functional activities. Remember, these are opportunities to unite your team, so be positive, motivate them, and when discussing the issues, make sure to focus on the facts of the issues, not the people involved. This helps minimize the politics.

For team members who have a lot of tasks or are tied to some key issues, check in daily with them to get updates and ensure that progress is being made on the top priority issues and tasks. For team members who have less tasks and who seem to be self-managed, check in less frequently.

Manage the Politics

As you move through the process, continue to *promote* your goal to SME's who are affected by the change. Make sure they understand the direction and rationale for the changes. They can then operate as internal promoters within their own functional area to further socialize these changes with others.

When creating the deck used to guide the monthly ESC meetings, always insert the project's charter at the beginning of the deck to remind people of the project's purpose at the beginning of every meeting.

> ## *Key Point*
>
> *Once your goal is clearly stated and you begin to form your project team, identify each team member's role in reaching this goal. This will create a highly motivated, highly efficient team and allow you to reach your goal quickly.*

Create Drilled-Down Financial Goals

As you saw when we defined our Goal, we have two main financial goals:

1. $1M in sales after one year
2. $5M in sales after three years

While these are good, it's important to work with your Finance team to drill down your year one goals into monthly goals to allow you to monitor the success of your product launch sooner (month by month) versus waiting for a year.

This is an exercise that is based on assumptions, so don't overthink it. Just meet with your finance team and work to document your best guess as to what your monthly goals in the first year should be and document the assumptions you are making. Remember, there will always be a ramp-up period and the degree of that ramp-up will depend upon a number of factors (how fast the roll-out occurs, how much advertising is used, if you're hitting the early adapters or a saturated market, etc.).

I've always found it helpful to identify the degree of confidence you have for each assumption since you'll probably feel pretty solid about some assumptions and aiming in the dark with other assumptions. That way, when you go back to review these assumptions after the launch, you can easily recall these details.

Now if your goals also include non-financial goals, such as ensuring customer service calls stay at .33/customer/month, then make sure to meet with the appropriate customer service lead before you launch to determine how this goal will be measured and how often it should be measured.

These monthly goals will be part of the pre- and post-launch measurements that we will discuss in Chapter 7.

5 Secrets to *Guide* a

Faster Product Launch:

1. **G**round Your Project with A Clear Goal

2. **U**nderstand the Required Changes

3. **I**ntroduce Tight Project Management

4. **D**rive a Data-Based Launch Decision

5. **E**stablish Pre- and Post-Launch Measurements

Chapter IV

The Second Secret:

Understand the Required Changes

Once you've properly grounded your project by articulating and socializing your goal, you'll next want to understand the changes needed to reach that goal. In other words, you'll need to start documenting your Business Requirements.

Say 'Business Requirements' to any marketing person, and you'll see eyes rolling and have people running out of the room. No one likes to document business requirements. It's a tedious process which forces you to make hard decisions sooner rather than later. But without a well thought out set of Business Requirements, your team will not be able to get their arms around the number of changes that need to happen in order to reach your overall goal.

How To Create Business Requirements

After years of being involved with product and marketing program launches, I've identified a best practice approach to documenting Business Requirements that will allow you and your teams to easily walk through this process and not continually have to update or go back and add a number of additional requirements.

I've named this best practices approach the Six Viewpoints Business Requirements because there are six different viewpoints that you will consider as you work with your team to define your business requirements. By viewing the necessary changes from these six different viewpoints, you begin to create a much more comprehensive set of business requirements, which will allow for a more thought out product development process and a product launch that includes all the necessary requirements.

Remember to be as specific as possible when describing your business requirements so that team members who are working within their own functional areas who have to address these requirements fully understand the specifics of your requests, are up to speed on the degree of necessary changes, and are aligned with the team's overall goal and the timing of this goal. Also, remember to state 'what' you need and stay away from a more technical view of 'how' it will be delivered.

In creating Business Requirements using the Six Viewpoints, walk through these six steps:

1. Identify the large changes first
2. For each of these large changes, look to see if there are underlying changes required in four main areas:
   a. Customer-facing business process changes
   b. Back-office business process changes
   c. Organizational changes
   d. Technology changes
3. Identify long lead-time items
4. Identify dependencies (things you need outside the scope of your team)
5. Identify interdependencies within the scope of your project team
6. Double check with the Murphy's Law exercise

If the scope of your project is small, run through this exercise quickly. If the scope of your project is large, take the time to thoroughly walk through these steps with your cross-functional team and you'll uncover all of the required changes, impacts and risks early on in the project.

Knowing which areas are impacted will be an evolving process, so again, expect changes to your business requirements. However, most of these answers become obvious once you have your goal clearly articulated and you gather input from your team. With these elements in place, the above exercise will go smoothly. Again, lean hard on your cross-functional team to identify the changes needed within each of their functional areas. If you need to walk through this exercise over several meetings, then document the business requirements in meeting #1 (for instance) and for meeting #2, have your team review them and be prepared to provide comments and updates.

It is of great importance that all of these changes be recorded in a "Business Requirements" document. This way you will keep the team on the same page and ensure that you are able to properly manage the required changes.

Some companies call the 'business requirements' document by another name (marketing requirements, product description, product requirements, etc.). No matter the name, the idea is the same... *document your vision and the supporting elements in detail.*

This document forms the foundation of your project plan. It also is a living document that your team members should use within each of their functional areas to further socialize the process and technology changes needed, and to help identify the personnel impacted by the changes.

Remember, this process is an interactive one and one that will continue to evolve as you move through the project. Let's use our Tracker product launch example to walk through the six viewpoints discussed above in more detail.

1. *Identify the Large Changes First*

Ask your team the following questions:

- What will be the main functionality of the software product (track expenses at a project level, be able to roll up expenses and drill down, security access, etc.)
- How the product will be manufactured
- How will it be packaged
- How will it be priced
- How it will be positioned and marketed
- How the product will be distributed and sold

- How the product will be tracked internally
- How will employees be trained
- What is the upsell path

2. *Identify Changes in Four Areas*

For Each of these Large Changes, ask your team if there are Underlying Changes required in these Four Areas:

1. Customer-facing business process changes

2. Back-office business process changes

3. Organizational changes

4. Technology changes

This step is where a deep dive into the nitty-gritty detail is needed. Make sure to take the time *now* to document at this level of detail because all required changes pop up at some time during your project. What you want is to cull out the changes needed in a detailed, systematic way *early on* in the project. Meet one-on-one with each team member to do the deep dives so you don't waste other team members' time, but get a commitment from them to review each other's areas for possible impacts to their own functional areas.

In our software product example, here is how we would answer the questions in this step (but would go into greater detail under each area when completing the business requirements document):

1. Which **customer-facing business processes** are affected and how are they affected
    a. Advertising
    b. Distribution
    c. Sales/purchasing
    d. Pricing
    e. Packaging
    f. Product
    g. Returns
    h. Referrals
    i. Customer feedback

2. Which **back-office facing business processes** are affected (manufacturing, account receivables, account payables, retailer pricing and contracts, budgets, product tracking, quality control) and how they are affected

3. Which **organizational changes** are needed (additional manufacturing headcount, training, adding customer service reps, adding sales reps). Not a lot of changes are needed here given that existing distribution channels and customer service teams are being used.

4. Which **technology changes** are needed (develop the new software product, develop additional system tracking for marketing, distribution and sales)

When filling in the Business Requirement document, start with the listed changes above and drill down with as much detail as possible. For instance, a customer-facing business process change is advertising. Detail out what type of advertising is needed (TV, print, radio, internet, brochures, data sheets). Again, remember to focus on 'what' you need versus 'how' the requirement will be met. In other words, don't let them become technical requirements.

3. Identify Long Lead-time Items

It is important to identify long lead-time items early on, understand the time required to complete them, know what other items are dependent upon them, and follow their progress closely as they can easily push back your project completion date if not properly managed. Of all the changes needed to reach your goal, only about 5% of them will be the long-lead time items.

In our software product example, the long-term items include:

- Developing the new software -- 120 days
- Finalizing product positioning and agreeing on advertising – 45 days
- Negotiating contracts with retailers – 90 days

Key Point

A key driver in determining how quickly you can reach your goal is the length of time it takes to complete your long lead-time items.

Let me hit on a side topic that is near and dear to my heart because I love to help companies create them, but it is also critical to managing a successful product launch – proper product positioning.

In John Coyne's book, *Opposite Marketing*, he highlights why core product positioning is so important. He states, "Creating a Core Positioning is critical to your marketing efforts for several reasons:

1. It forms the foundation for developing and marketing your product or services.

2. It brings focus to your marketing and sales efforts and insures that all of these efforts have continuity and singleness of purpose.

3. It clarifies for your ideal client exactly who you are, why you exist, and how you are different from other independent professionals.

4. It states the reason for your brand's existence and, once successfully established, it should rarely be changed."

It's been my experience that creating your product's core positioning takes much longer, is much harder, and involves many more people than you'd initially anticipate. So plan for this important task appropriately, and work with your marketing project team members to ensure that movement continues to be made every week given the dependence the above items have on this output.

4. Identify Dependencies (things you need outside the scope of your team)

In our software product example, the dependencies include:

- Internal resources will be made available for this launch
- Budgets will be made available for this launch (as stated in approved business plan)
- External resources (the software testing team) will be available when needed

5. Identify Interdependencies within the scope of your team

In our software product example, the interdependencies include (there will be many more):

- Before we can develop the software, we need to close on business requirements
- Before we can create advertising, we need to finalize the product positioning

Listed below is a section of the Business Requirements document created to roll out our new software product. Keep in mind that as you begin to collect these requirements, you'll notice that some of them need to be further detailed out to adequately capture all of the underlying changes, while others can be left at a greater level of detail. How will you know? As you begin to write down the detailed changes, you will exhaust the question of 'are their supporting changes needed?'

Here is an example of a Business Requirements Document.

Business Requirements Document (partial view):

	Requirement	Functional Impact	Decisions/ Assumptions
1.1	New software that tracks expenses at a project level	**Engineering**: software development **Quality Assurance**: set up new processes to track defects	Customers will not need to track more than 5,000 projects' expenses
1.2	Eye catching packaging	**Marketing**: design new packaging; user test with a small sample of our target market	Packaging will reflect agreed upon product positioning

Some companies use another format for collecting business requirements. Not a problem. Just remember to document what you need in detail.

It is inevitable that additional Business Requirements will be identified as you move through the project -- expect this, and make sure your working level team expects this. What needs to be emphasized here is to take the time to map out a detailed Business Requirements document early on in the project to build a better foundation on which to execute your product vision.

Key Point

*It's important to identify as many of the required changes as you can **early on** in the process versus later down the road.*

6. Conduct the Murphy's Law exercise

This is an exercise that I use to help our project team to think outside of the box and have a little fun at the same time.

Have your team think of the worst possible scenarios and make a list of everything that can go wrong with your project (it's actually a fun thing to do… adds levity to it all).

Put a mark next to those that have a high probability of happening. For those high probability items, identify how the team can keep those things from happening. If appropriate, work these solutions into the project plan.

Create an Overall Business Model

As you continue to add requirements, always go back and update your Business Requirements document and communicate changes to those identified in the Functional Impact column. To help with the process of understanding what each Business Requirement affects, I like to create an Overall Business Model. I have found this to be so helpful that I now consider it a Best Practice because it helps with communication in so many functional areas.

What is an Overall Business Model? It is a one-page document that graphically shows the entire business in three buckets:

1. Customer-Facing Business Processes
2. Your Product(s)
3. Back-Office Facing Business Processes

By going through the exercise of defining your business on one piece of paper, it allows for great discussion and brings divergent views for others to think about. At the end of the exercise, you will have created *a common view* of your business that you and all the members of the Governance group (ESC, project team, subject matter experts) can use to communicate in a standard way about your business.

I was helping a client get their arms around the prioritization process they used for system enhancements, and to help guide our discussions, I created a business model for their company. What became immediately apparent was that it was so much easier to describe what part of the business the enhancement would affect once they had a common business model. It simplified their communication and created a common language for everyone to use.

I have used our software example to show their overall Business Model on the next page:

Typical Business Model

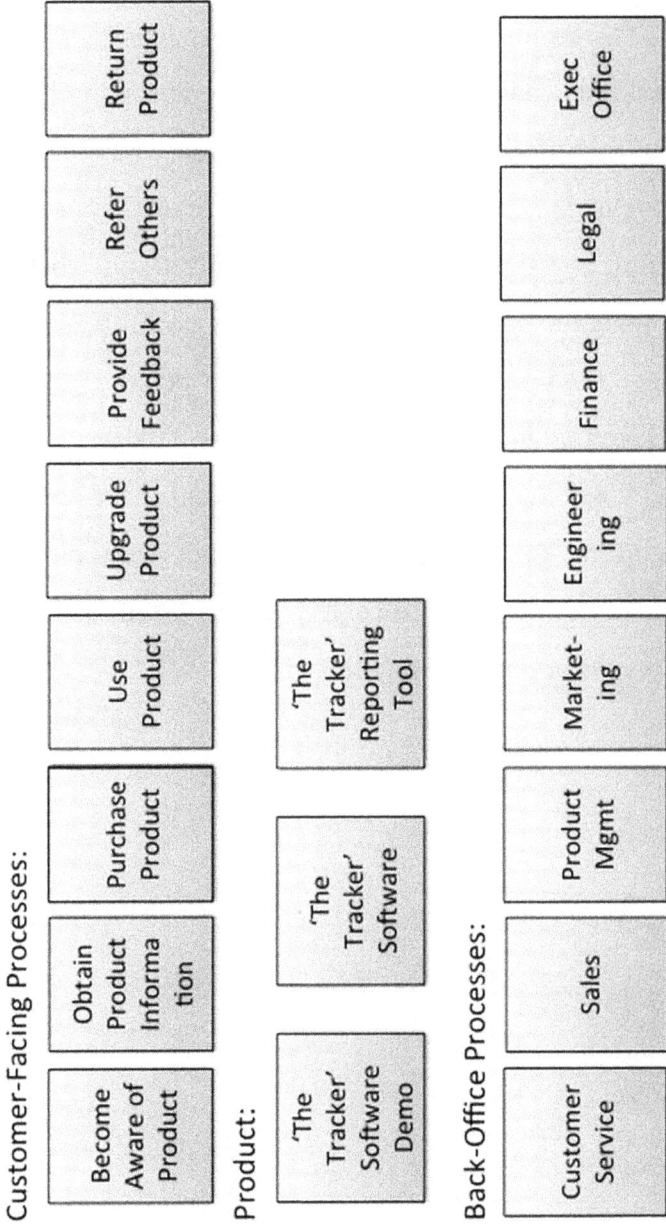

Customer-Facing Processes:

Become Aware of Product	Obtain Product Information	Purchase Product	Use Product	Upgrade Product	Provide Feedback	Refer Others	Return Product

Product:

'The Tracker' Software Demo	'The Tracker' Software	'The Tracker' Reporting Tool

Back-Office Processes:

Customer Service	Sales	Product Mgmt	Market-ing	Engineer-ing	Finance	Legal	Exec Office

Manage the Politics

After identifying your key dependencies, work with those that are accountable for these deliverables and bring them into the fold of your team. They don't need to attend all the team meetings, but they do need to feel that they are a part of the team, that their input has been received, and that their actions impact the success or failure of this team. If necessary, pull in their boss and their boss' boss in a positive way to highlight the importance of these deliverables. Update your boss on the progress of these dependencies and have them positively reinforce the need for their on-time delivery during the course of other meetings since it will have a direct impact on the success of your product launch.

5 Secrets to *Guide* a

Faster Product Launch:

1. **G***round* Your Project with A Clear Goal

2. **U***nderstand* the Required Changes

3. **I***ntroduce* Tight Project Management

4. **D***rive* a Data-Based Launch Decision

5. **E***stablish* Pre- and Post-Launch Measurements

Chapter V

The Third Secret:

Introduce Tight Project Management

So far, we've covered two Secrets:

- First Secret: Ground Your Project with A Clear Goal

- Second Secret: Understand the Required Changes (Business Requirements)

Now, we're ready to cover the next step:

- Third Secret: Introduce Tight Project Management

This is where the term "the devil is in the details" comes into play. Yes, I have seen people implement changes without a lot of detailed plans, without documenting their Business Requirements, and without a lot of communication. But this has *only* been successful when one or two people are affected by the change.

When a change affects a number of people or a number of functional areas (which most changes do), then systematic management of the change is required.

It has been my experience that the optimal way to systematically manage these changes is through

- Seven Best Practice **Techniques** to Manage Your Project Team, and
- Seven Best Practice **Tools** to Manage the Work

More on both of these topics below.

This step in the process (Tightly Project Managing the Changes) will take you the longest and will try your patience the most. It may even make you want to quit. However, stick to it because it is also the most impactful step to *making the change happen.*

Looking back in my career, I have always added the most value when I successfully implemented a change within an organization. What I've come to realize is that, for most people, understanding how to implement a change does not come naturally. And wanting to change is not a choice most people would make.

Change is not only hard for those leading the change, it is also hard for those impacted by the change. Most people like things to stay the way they are because that's within their comfort zone.

I have found that there are Seven Best Practice **Techniques** that are common to successfully **Manage your Project Team**:

1. Cross-functional work

2. Full disclosure of information

3. Accountability

4. Tight Project Management of the Changes (see Tools list below)

5. Weekly Project Team and Monthly ESC Meetings

6. Supportive Leadership

7. Constant Prioritization

There are also Seven Best Practice **Tools** that are needed to properly **Manage the Work**. The degree to which a change affects multiple areas is the degree to which you should utilize these Tools:

1. Team Charter
2. Launch Criteria
3. Project Plans
4. Issues Log and Action Items Log
5. Jeopardy Notice
6. Succinct Weekly Project Team Meeting Deck
7. Succinct Monthly ESC Meeting Deck

If you use nothing else, these will help keep your project moving and on track, and will help keep upper management abreast of progress and connected on key issues that require their involvement, again, minimizing the politics.

In the pages that follow, I will go into detail about both the Best Practice Techniques for Launch Team Management and the Best Practice Tools needed to successfully manage all the work required to implement your product launch.

Seven Best Practice Techniques to Manage Your Project Team

1. Cross-Functional Work

You cannot have people who are involved in implementing a change work in silos. IT JUST DOES NOT WORK! I cannot stress this enough.

What does cross-functional work look like?

- Cross-functional project teams

- Cross-functional ESCs

- Cross-functional subject matter experts

- Cross-functional project management (project planning, issue tracking, escalation and resolution)

- Clear roles and responsibilities between each of the functional areas

- Collaborative teaming

- Individual accountability for their work (team members, ESC, subject matter experts, vendors, partners, etc.)

When I first came on board with one project, it became apparent that there were several legal changes that would be required, yet no legal counsel had been involved because they didn't want to waste their lawyer's time. Although it did not make sense for their lawyer to sit in on every team meeting, it was important that he understood the overall goal we were working on, and the legal work that was required to reach this goal. So, I set up a meeting to review these with their lawyer and obtained dates for when his pieces of the project plan would be complete. He then understood the overall goal, his role within reaching this goal, and the consequences of not meeting his dates.

Key Point

The majority of product launches require changes that affect several functional areas, so you need to project manage from a cross-functional perspective. If there is a department that's hard to work with and you avoid involving them, it will become a bigger problem for you down the road.

2. Full Disclosure of Information

To properly manage changes, you need to *expect* to have issues, and you need to train your project team members and ESC to expect to have issues. To that same degree, you need to set the expectation that your team (and the ESC) is accountable to close these issues and they are equally responsible for identifying issues *early on* in the project.

I mentioned above that the majority of changes required to meet your goal will have cross-functional impacts, that's also true for the majority of issues – most issues raised in projects will have cross-functional impacts. So, to operate on a need-to-know basis when discussing how to handle issues in essence hides additional issues that will eventually be uncovered down the road. This lack of disclosure also instills distrust amongst the team members, which directly affects productivity.

While working as a consultant on a product launch team, members within the IT department operated mainly within their own department. They didn't see the value of communicating with other departments when it came to documenting requirements or making trade offs. When they uncovered issues, they tried to solve them on their own and didn't communicate their solutions to other departments. However, the impact of the issues directly related to other functional areas.

They had good intentions (solve the problem as fast as they could), but disastrous results (they created a very inflexible system and team members did not trust them).

> ### *Key Point*
> *Create a team environment where identification of issues is a good thing and even better when done early in the project and on a cross-functional basis.*

3. Accountability

You need to set the expectation that team members are accountable for:

- REALLY meeting their deadlines
- Identifying issues early on in the project
- Actively participating in cross-functional meetings
- Creating, updating and managing their tasks on the project plan

- Quickly closing issues/action items assigned to them
- Sharing information – full disclosure
- Building a better team environment for all to work
- Setting ground rules together for how the team will operate and communicate

We've all learned that if you don't have buy in from the top, you won't go anywhere. The same holds true for making changes. Unless you have buy-in from those affected, and accountability that they'll support the change in each of their functional areas, your change won't happen.

I had a consulting assignment to help launch a new consumer product. We formed a cross-functional team and had clearly defined each of our roles and responsibilities. However, one of the key influencers began to participate less and less until I pulled him aside and talked about the need for his involvement. He didn't like to attend team meetings. So in the end, I had to set up specific meetings to keep him involved (which was redundant for me), but when we needed his input and power to address issues, he was available to us.

Along with setting clear goals, let team members know the impact of their work and provide rewards along the way for key milestones and the tougher project tasks, along with a team reward for accomplishing the overall team goal.

Key Point

Set expectations about how your team will operate ahead of time by agreeing, as a team, to ground rules. These might include being on time for meetings, actively participating in discussions, taking ownership of your area, cell phones on vibrate and no texting, focus on solving problems vs. pointing fingers, etc.

4. Tight Project Management of the Changes

I always see things in 'buckets.' It helps me organize and compartmentalize all the various pieces of work that need to be managed. That's why I see two main buckets when it comes to change management:

- What is the Work? (what are the required changes)
- How will we get the work done? (the project plan)

Now, I'm not saying that its fun to put a project plan together. But I will say that if you take the time to incorporate all the required changes into one document (business requirements), then put all these changes into phases and connect them to other dependencies (project plan), then you will be MUCH better off in implementing all of the required changes.

Creating the project plan is one step, but the work does not end there. You need to use your project plan as your bible of all the moving pieces of work. You will need to:

- Continually update your project plan with the most recent information about the changes required, functional areas impacted and the timing of these new changes

- Use the project plan when you meet with your functional leads to check the progress in each functional area and to make sure people are following through on their individual tasks
- Use the project plan to identify impacts when things change (and they change every day)
- Use the project plan to steer your team each week to highlight upcoming milestones and to celebrate achieved milestones
- Use the project plan as a communication vehicle so each functional area understands their impact on other functional areas

This will be your key document and a way to simply collect the abundance of information that comes up in your daily discussions. It is also a way to track the key dependencies that your project relies on to achieve your goals. I use the project plan to remind me to continually watch the progress of these items and ensure that they are available when they are needed.

Key Point

There should be 2 project plans:

*1) A very **detailed**, day-to-day view that covers the changes across all functional areas, and*

*2) A **high-level**, one page project plan to use with your Executive Steering Committee or to get a quick view of the overall project.*

For those that are newer to detailed project plans, I have provided a starter list of what you may want to include in your project plan in Appendix B.

5. *Weekly Project Team and Monthly Executive Steering Committee Meetings*

After years of launching products, I've tried a number of different ways to manage the working level project team, and the best practice that has come to light is to have very tightly managed weekly project team meetings.

At the start of every meeting, I quickly review the Team Charter and one-page project plan to reorient everyone as to the overall plan. We then review What's Changed Since We Last Met and Key Issues to get to the hottest items first and have detailed discussions about the specific next steps required ensuring movement and closure of these issues. A template of a Weekly Project Team Meetings Deck is provided in Appendix C.

For issues that are spinning out of control or are just not moving forward and are putting your launch in jeopardy, I use a Jeopardy Notice. This is a document that you use to communicate the degree of the issue (Red, Yellow or Green), and to spell out in detail the background, the issue, the implications, identify who owns the issue, identify what help we need from whom and by when (step by step), and a date by when we expect the issue to close. More on Jeopardy Notices under Best Practice Tool #5 in the next section.

I've also found that a best practice is to meet with your ESC on a monthly basis. This allows affected executives to stay up to date on the progress of the product launch, highlights what the team needs from their functional areas, and is a forum for direct communication with people who can help clear obstacles. This also allows you the opportunity to remind people of the importance of your project and gives them the knowledge to speak to others about the product launch, further promoting it within the company.

Key Point

If you want to stay connected to your team members, lead tightly managed weekly team meetings where most of the time is spent brainstorming how to close the top issues and a small amount of time is spent on what have we done and what have we got to do in the next few weeks.

6. Supportive Leadership

Teams can do amazing things when their leaders give them a well-articulated, clear goal, identify team members' specific roles, continually reinforce the importance of reaching this goal, and give appropriate rewards (meeting key milestones, thinking outside the box, etc.). Great leaders provide motivation and expand the boundaries under which team members operate to allow them to excel in their roles.

Leadership comes into play at 2 levels: the project team level and steering committee level.

The project team leader needs to lead and reward at a tactical level. They must have the backbone to identify the hard issues and work with others at higher levels to discuss and address them. They must be passionate about reaching their goal, and convey that every team member plays a critical role in reaching this goal. They must be strong while being diplomatic. They must be serious while still bringing fun into the everyday work. They must be willing to back their teammates up 100% and take responsibility and action when mistakes happen.

The leaders in the ESC must operate at a more strategic level, making sure to aspire greatness from their team members and making the project team feel confident that issues raised to them will be heard and managed to closure, without retribution.

"The leaders who work most effectively, it seems to me, never say "I." And that's not because they have trained themselves not to say "I." They don't think "I." They think "we"; they think "team." They understand their job to be to make the team function. They accept responsibility and don't sidestep it, but "we" gets the credit... This is what creates trust, what enables you to get the task done."

Peter Drucker

7. Constant Prioritization

You can spend your entire day working, but unless you are working on the right things, you'll be wasting time. When projects have just hummed along, I've noticed that I had continually checked what I (and others on the team) was doing against the project goals to make sure I wasn't wasting time, and to make sure that I was working on the most important things first.

Every morning, I'd sit down and make a list of the things I just had to get done that day, along with checking the project plan and issues log to see what I needed to do to advance them to closure. This exercise kept me on track to make sure I was always focused on the most important, time-critical things. Once they were done, I could focus on less time-sensitive things and move them along. Again, it goes along with the theme of this book, unless it's managed, time slips away and you don't accomplish what you set out to do.

As things came up during the day, I went through the same exercise… how does this fall in the priority of what I have to get done today? If it's lower, then I'd usually put something on the calendar for the next day or two to make sure that we'd have time to address the issue or discussion.

In David Allen's book, *Getting Things Done: the Art of Stress-Free Productivity*, he has found that, "In training and coaching thousands of professionals, I have found that lack of time is not the major issue for them (although they themselves may think it is); the real problem is a lack of clarity and definition about what a project really is, and what the associated next-action steps required are. Clarifying things on the front end, when they first appear on the radar, rather than on the back end, after trouble has developed, allows people to reap the benefits of managing action." This reinforces the need to document your team's goal up front and create a detailed project plan that guides and pushes the team on a day-to-day basis.

One company that is particularly good at prioritization is Apple. In Walter Isaacson's book, *Steve Jobs*, Tim Cook is quoted talking about Jobs, saying that, "There is no one better at turning off the noise that is going on around him. That allows him to focus on a few things and say no to many things. Few people are really good at that." Steve Jobs used his Monday morning executive meetings to constantly reinforce the two or three priorities that he had for the company, which in turn allowed his executive team to narrow down their focus and say no to the noise.

Seven Best Practice Tools to Manage 'the Work'

I have used these tools over and over and I continue to be amazed at how well they work to keep my projects on track, along with using the above Best Practice Techniques for Project Team Management. The reason these tools work is that they're simple, easy to understand documents that allow people to quickly take in information and easily provide updates back to you to pass on to others that are affected.

With that said, here is my list of **Seven Best Practice Tools** to manage the work:

1. Team Charter
2. Launch Criteria
3. Project Plans
4. Issues Log and Action Items Log
5. Jeopardy Notice
6. Succinct Weekly Project Team Meeting Deck
7. Succinct Monthly ESC Meeting Deck

1. Team Charter

Why do you need a team charter when you already have a clearly articulated goal? Well, they're usually pretty similar, but it specifically states **how** this team will work together to achieve that goal. It also fills in the gaps with other pieces of information that help ground the team around a common set of assumptions.

Work with your team to create your team charter, and include:

- Overall Goal (covered earlier)
- Role of Team in meeting this goal
- Phases (if any)
- Scope (what's included, what's not included)
- Key assumptions, constraints, risks, and dependencies
- Roles and responsibilities of each team member

Here is a Good Example of a Team Charter:

Overall Goal:

Ensure that Software for Business Inc. launches a new software package called 'The Tracker' on October 15, 2012 (six months from now) in order to capture the untapped market of small and medium businesses that are not able to easily track their expenses without purchasing very expensive, large-scale software programs. By leading the launch of this product, we will enable our company to capture $1M in incremental revenue within a year, and $5M within three years.

Role of Team:

The team plays the following role:

- Document business requirements
- Engineer and build product and packaging
- Distribute to distribution channels
- Promote product
- Build product roadmap to reach 3 year revenue goal

What's In Scope:

- Customer Facing: Target audience awareness, customer purchase, customer use, customer support, advertising, customer feedback

- Back-Office Facing: pricing, engineering, testing, distribution, product management, marketing management, business planning, budgets, legal, finance (billing and receivables), customer service

What's Out of Scope:

- Customer Facing: other product advertising

- Back-Office Facing: other product lines

Key Assumptions/Constraints/Risks/Dependencies:

- Team will get proper staffing & funding
- IT will be able to meet software development changes on schedule
- Vendors will be able to supply needed components
- Product will be distributed through existing distribution channels
- Product will be priced at a mid-market price
- Product will be marketed as all new products are marketed, though targeted only to small and medium businesses.
- Competitor will be launching their product next year
- Target audience has a strong need for this type of software product

Role of Project Team Members:

Project Manager	Joe Barnister
Marketing	Taryn Robins
Sales	Grant Davis
Engineering	Nicole Johnson
Legal	Dave Montana
Finance	Kerry Clark
Customer Svc	Frank Ravens

Here is a Bad Example of a Team Charter:

Role of the team is to meet the stated goal.

2. Launch Criteria

Once you have properly documented your Team Charter, you should create Go / No-Go Launch Criteria. You will use these criteria near the end of the project to assess your launch readiness.

When creating Launch Criteria, think of them in two different buckets:

1. Customer-facing launch criteria
2. Back-office launch criteria

The purpose of Customer-facing launch criteria are:

- Align cross-functional vision of what is required, from the customer's perspective, to successfully launch
- Use as basis for key decisions, such as to prioritize bugs found during testing; determine scope of pre-launch user trials; and/or systematically assess your launch readiness

> ## *Key Point*
>
> *It's important to have a solid product positioning statement that has been documented, socialized and agreed upon by the appropriate parties. If you have this, you can create tasks in the project plan that 'operationalize' your product positioning. You can also craft your launch criteria to include the requirement that the product launch properly reflects the agreed-upon product positioning.*

In our paper example, the software product positioning is: Affordable, reliable, timesaving expense tracking for small and medium businesses.

To explain this further, we'll create launch criteria using our software product example.

Customer-Facing Launch Criteria:

- New software product is available in a pleasing, eye-catching package that reflects the product positioning
- Distribution channels are trained and new software is on the shelves
- New software is available on our website to view, obtain information, and buy and it properly reflects the product positioning
- Customer is able to easily locate our new software within affiliate stores and their websites
- Customer is able to easily purchase our new software
- Website homepage highlights the new software and pulls in new, qualified leads for our distribution teams
- Customer service is staffed and ready to take calls, chats, emails
- Customer is able to receive software within one business day, once ordered via our website
- Customer is able to easily return software, if necessary

Back-Office Facing Launch Criteria:

- We are able to engineer new software with existing technology and manpower

- We are able to take orders from distributors
- We are able to distribute to appropriate distribution channels for sale to end user
- We are able to track orders, sales, returns, defects, customer service calls through our existing financial systems
- We are able to provide CEO and senior management team with summary reports each month in same format as other product tracking reports

One key assumption across all criteria is that these elements are done accurately, on time, with quality, and within our cost and launch timing assumptions.

During the course of the launch preparations, we continuously use these Launch Criteria as a way to measure progress and prioritize issues. It is also important to double-check your business requirements and project plan to ensure that everything needed to support these launch criteria has been included.

Two weeks prior to launch, as assessment will be conducted walking through each criteria and identifying what is OK and not OK for launch. We will also identify the associated risk (high, med, low) with launching given the assessment of each criterion. This is how you will get a data-based go / no-go launch decision. More on this topic to follow in the next chapter.

3. Project Plans

Depending upon your expertise and your teams' expertise, create your project plan with a tool that everyone can understand. I've used Microsoft Project when dealing with mostly technical people and Excel when dealing with a mix of cross-functional team members.

How do you start to create this plan? The plan should cover all the changes identified by the team in your Business Requirements document. It should also identify and track all dependencies and interdependencies. So, start with that list and then break all the changes down into milestones first, and then add detailed tasks below each milestone, making sure to identify if each task is dependent upon another task.

Here is a list of possible areas to cover in your project plan (a more detailed list is provided for you in Appendix B for you to use when creating your detailed project plan):

- Strategic Foundation ('fit', project goal, ESC)

- Marketing (product, pricing, distribution, promotion)

- Finance (GL tracking, funding)

- Legal (contracts, legalities, Terms & Conditions)

- Engineering (system changes, testing)

- Billing (billing and collections)

- Customer Service (in-person calls/chats, automated calls, auto web)

- Impact on the Customer (current and future)

- IT (system impacts)

- Steering Committee socialization

- Outside vendors/partners contributions

- Key dependencies

A more inclusive list of possible project plan tasks is provided in Appendix B. Keep in mind that this is just one version of a plan and that your version might be much shorter or longer depending upon the scope of your changes. I do suggest always starting out with the more strategic tasks and then working your way down to the more tactical tasks. This is an iterative process in that you start to build your plan based upon what you think the changes will be, but as you get further into your project, additional changes might be identified, or the need to drill down much deeper on an existing change might be necessary to ensure that the change gets completed on time.

My project plan was always my bible. I read it and re-read it, updated it constantly, got to know all dependencies by heart and memorized all the interdependencies. This allowed me to quickly understand the impact when changes arose.

I brought it with me wherever I went (along with the Issues/Action Item Log and the one-page project plan) so I always had access to the latest information in case I ran into team members in the hallway or in other meetings and I could quickly get updates from them on the fly and have all of the tasks and due dates at my fingertips. I could also use the logs to remind me of the hottest issues that need to be constantly guided to closure.

> ### *Key Point*
>
> *I set up weekly one-on-ones with each functional lead to ensure we had regular drill-down sessions on their tasks and key dependencies. This was my time to really understand a specific area, or put pressure on them when needed.*

I used the detailed project plan to prepare for my one-on-ones with the functional leads, and to prepare for our weekly Project Team meetings. Because I never went anywhere without our project plan, people knew that I'd always be following up with them on their tasks and the issues that they were helping to close.

As stated earlier, there should be 2 project plans: a very detailed, day-to-day view that covers the changes across all functional areas and a high-level, one page project plan to use with the ESC and in our weekly team meetings. This will allow you to maintain both a tactical view for managing all the work on a day-to-day basis and a strategic view for managing at a higher level to ensure you see the bigger picture.

Creating a project plan from scratch takes a lot of effort, so I provided a template for you in Appendix B.

Here you'll find an outline of what I use as my starter set of tasks and milestones. You can look through this list and see which of these apply to you to make your own detailed project plan, and make sure to add your own tasks that are not listed in Appendix B but apply to your project.

On the next page you will find an example of a high-level project plan. I use this to keep my head out of the details when talking to executives or to quickly communicate the bigger picture. It was also helpful when large-scale changes were happening and I needed to quickly understand the implications on the other sections of the plan.

Here is an example of a high-level, one-page project plan:

One-Page Project Plan

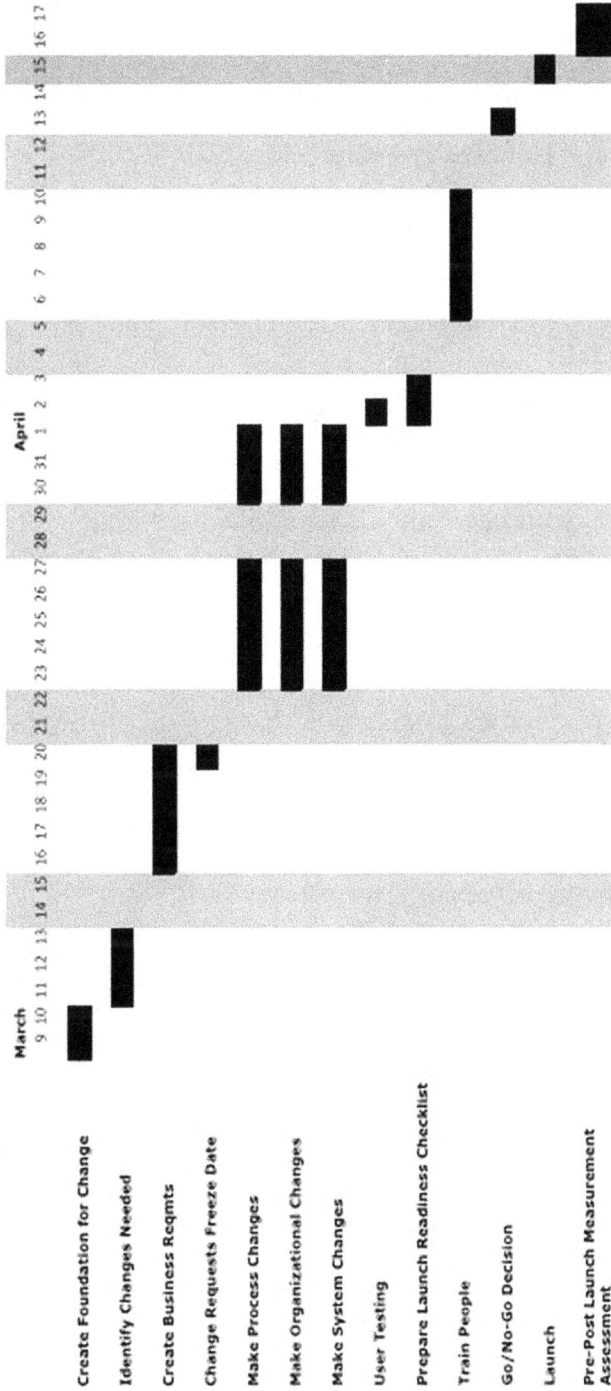

March / April Gantt chart with the following tasks:
- Create Foundation for Change
- Identify Changes Needed
- Create Business Reqmts
- Change Requests Freeze Date
- Make Process Changes
- Make Organizational Changes
- Make System Changes
- User Testing
- Prepare Launch Readiness Checklist
- Train People
- Go/No-Go Decision
- Launch
- Pre-Post Launch Measurement Assessment

4. Issues & Action Items Log

Again, I never went *anywhere* without these 3 documents:

1. Detailed Project Plan

2. Issues and Action Items Log, and

3. The High-level, One-page Project Plan

They were stapled together and continually updated each time I spoke to different team members throughout the day. These documents allowed me to work with the team to identify issues, quickly understand the impact, prioritize them, and document the path to close them.

The Issues Log easily shows the top issues in order of importance and quickly guides me to where I should be paying attention on a given day. Included in the Issues Log are the following pieces of information:

- Description of the Issue
- Responsible person's name (Lead)
- Priority of the Issue (High, Medium, Low) – issues were listed in the log based on their priority, High's were always listed first
- Status (Green, Yellow or Red)
- Updates (date each update)

Of all the documents listed above, you will be making the most updates to the Issues Log given the tactical nature of managing the details of a product launch. By continuously updating this document, you can quickly and easily manage the details of the day to day work. Also, by letting people know that you are tracking their issues and will be following up with them, you encourage them to address them and move them to closure.

For simplicity, this is also where I collect Action Items. I differentiate the two by thinking of action items as items that are smaller in scope and haven't been captured in the project plan. It is an easy way to collect updates on all project related items and not continually update the detailed project plan with nitty-gritty things that are usually closed out within a week.

Here is an example of an Issues Log and Action Items Log:

Issues Log

#	Issue	Lead	Priority	Status	Updates
1	Engineering: Unable to secure the engineering resources required	Henry	High	Yellow	5/12/12: Henry spoke to his boss who will re-prioritize work in order to make resources available within 3 days. 5/10/12: New issue.
2	Marketing: New packaging vendor is unable to work on prototype of packaging to meet our dates.	Nicole	Medium	Green	5/10/12: Nicole speaking with our current vendor who most likely will be able to make our dates if we move quickly. 5/8/12: New issue.

Action Item Log

(Note: these are items that are smaller in scope and haven't been captured in the project plan. This is just an easy way to collect updates on project related items and not continually update the detailed project plan with nitty-gritty things that are usually closed within a week.)

#:	Action Item	Project/ Lead:	Priority	Status	Updates
1	Obtain new photos for website content from our ad agency	Elizabeth	Medium	Green	4/28/12: Eliz to provide by 5/20/12

Key Point

Every piece of information I collected

about my project (no matter how big

or small) went into one of these

documents:

- *Team Charter*

- *Project Plan*

- *Issues Log*

- *Action Items Log*

This helped ensure that information

did not get lost in the shuffle (because

there are a lot of moving pieces) and

allowed me to quickly understand the

impact of new information.

5. *Jeopardy Notice*

Any issue that is launch-critical and has a status of Yellow or Red needs to be tightly managed to ensure timely closure. One tool to help with this is a Jeopardy Notice.

A jeopardy notice is a one-page document that clearly states the issue and details out the steps being taken to quickly close this issue (every action item lists the specifics of what will be done, a specific due date and a specific name(s) of who is going to do what by when).

The Jeopardy Notice should be updated daily to make sure that progress is being made every day, and it needs to be routed to those affected by the issue, and those who are able to help close the issue.

This is a helpful drill-down document that prompts you to dig deeper on key issues to assure closure quickly. It is also a helpful way of quickly updating those not previously involved. All Jeopardy Notices are distributed to the ESC to keep them informed of the hottest issues being worked and to highlight what specific action may be needed from a specific ESC member. A template of a Jeopardy Notice is provided later in this chapter.

On one assignment, I asked a team member to complete a Jeopardy Notice on an issue that needed additional attention or it would derail the entire project. The person was against filling out the notice because it was 'just another document.' Once he completed it, however, it became immediately clear to him that he needed help from a senior manager in another functional area. The document was an easy way to bring this senior manager up to speed and spell out exactly what we needed from him and by when. By the next day, the problem was on a corrective path and the project was back on track. If we had delayed writing up the notice, we would not have gotten the right people on the problem in time.

Set the tone early with your team that bringing up issues is the right thing to do, and that significant issues need to be tightly managed to ensure that they are closed quickly.

Train your team to use a Jeopardy Notice (see next page) when needed, to ensure that the issue is quickly communicated, that those responsible understand the steps they need to take to properly manage the issue, and those above them are watching to ensure that it is closed quickly.

Here is an example of a Jeopardy Notice:

Software for Business Inc.
Jeopardy Notice as of (date)

Project: Software Launch of 'The Tracker'	Jeopardy Owner: Henry R
Jeopardy Orig Date: 5/21/12	Jeopardy Status (G/Y/R): Red

Milestone/Deliverable in Jeopardy:
- Completed software from engineering (pre-test)
- Testing of this software

Issue & Impact:
Our Engineering team is understaffed due to unexpected resource departures and they are now unable to provide the completed software deliverables on time; also, the testing environment is not always available when needed. This directly impacts our ability to launch.

Recommended Plan of Action & History:

#	Action Item/Due Date	Lead	Priority	Status
1	Meet with Sam, VP-Engineering to see if work can be re-prioritized by 5/22	Henry	High	5/22/12: had mtg; result is Sam to re-prioritize and then meet back at 5pm EST today. 5/21/12: scheduled mtg for 5/22/12 at 11:30am EST with Sam & Henry.
2	Meet with Director of Testing to free up testing environment by 5/22	Cheryl	Medium	5/22/12: testing environment was available today; testing highest priority possible causes. 5/21/12: Escalated issue to Jane, Director of Testing; will have limited access each morning this week from 8am-10am EST.

Jeopardy Status Definitions:
Green with Issues: If not resolved, could risk a milestone, deliverable or project cost
Yellow: If not resolved, will risk a milestone, deliverable or project cost
Red: If not resolved immediately, will risk a major milestone, deliverable or project cost

Distribution Procedure -- If the Jeopardy is:
Green with Issues: Distribute to Project Team
Yellow: Distribute to Project Team, their Direct Supervisors, and Project Sponsor
Red: Distribute to the Yellow Group and add Executive Steering Committee members from the affected functional groups.

6. *Succinct Weekly Project Team Meeting Deck*

Rather than walking through the detailed project plans each week with our cross-functional team, I'd pull out the key things we had to review:

1. Team Charter to reorient everyone
2. One-page project plan to put the detailed project plan into a snapshot view
3. What's Changed Since We Last Met (to keep everyone on the same page and check to see who might be affected by the new change; example: we now won't be taking live customer service calls, only live chats and emails)
4. Overall Project Status
5. Key Issues (based on issues log, pull out only the high priority issues and issues that are Red or Yellow)
6. Completed & Upcoming Milestones
7. Action Items (ongoing list from meeting to meeting to capture the miscellaneous things; don't always have time to cover this in the meeting)

A template of a best practice Weekly Project Team Meeting Deck is provided in Appendix C.

7. Succinct Monthly Executive Steering Committee Meeting Deck

Your ESC plays a very powerful role in your larger team, if managed properly. They can add strategic insight, promote the importance of your team's goal, clear paths at executive levels, and work to help close issues that might hinder your team's success. So, use this resource wisely, let them know their intended role, and get buy-in from them.

In order to keep them engaged, you will need to keep them up to date on your project's status and the value their resources are providing. Given the level of people in the ESC, you'll want to take as little of their time as possible.

I recommend a monthly ESC meeting. In this meeting, you want to review the same items that you review with your working-level team, but take the details up several levels and concentrate on what you need from these executives to make your team successful.

In your monthly ESC updates, include:

1. Team charter

2. One-page project plan (to reorient everyone)

3. Key Issues (calling out where you specifically need their help)

4. Completed & Upcoming Milestones (to highlight accomplishments and keep them up-to-date on upcoming work)

Key Point

Remember, you formed the Executive Steering Committee so they could help you. Don't be afraid to use them and have the mindset that their job is to help you and your team to succeed. Far too often, people shy away from using this powerful resource because they don't have the confidence to engage with senior management. Step up to the plate, speak to them in their language and utilize them to help you when needed.

5 Secrets to *Guide* a

Faster Product Launch:

1. **G***round* Your Project with A Clear Goal

2. **U***nderstand* the Required Changes

3. **I***ntroduce* Tight Project Management

4. **D***rive* a Data-Based Launch Decision

5. **E***stablish* Pre- and Post-Launch Measurements

Chapter VI

The Fourth Secret:

Drive a Data-Based Launch Decision

So far, we've covered three Secrets:

- First Secret: Ground Your Project with A Clear Goal

- Second Secret: Understand the Required Changes (Business Requirements)

- Third Secret: Introduce Tight Project Management

Now, we're ready to cover the next step:

- Fourth Secret: Drive a Data-Based Launch Decision

So, you're ready to launch... right? Not so fast. Things are never black and white, they are variations of gray.

Three main things need to happen at this point:

1. You and your team need to do an honest **assessment of your launch readiness** (using the pre-defined launch criteria)

2. You and your team must decide on a **recommendation** for the ESC, to launch or not to launch, based upon the launch readiness assessment

3. This launch assessment and the project team's recommendation must be properly communicated to your ESC to allow them to make an **informed launch decision**

A formal go / no-go launch decision is extremely valuable and adequately prepares your organization for the new product launch. This decision also allows the senior executives (the ESC) to be fully aware of the nuances of the project and the risk levels involved in launching.

Assess Your Launch Readiness

To properly assess your launch readiness, work with your project team to complete a detailed Launch Readiness Assessment (example provided later in this chapter), which directly mirrors your pre-defined Launch Criteria, but with more detail. This assessment should list each of the launch criteria elements and for each one show three things:

- Percent Complete
- Risk in launching (high, med, low)
- Person responsible

With this detailed view, you are able to get a detailed, data-based assessment of your launch readiness and see which areas might be proving to be a higher level of risk than anticipated.

Prepare a Project Team Go / No-Go Launch Recommendation

Work with your project team to properly prepare a Go / No-Go Launch Recommendation. Do this by walking through the completed Launch Readiness Checklist, getting buy-in from the team on what's been documented. At the end of this review, discuss and agree upon a launch recommendation for the ESC (see example at the end of this chapter).

Guide the Executive Steering Committee to Make an Informed Go / No-Go Launch Decision

To properly guide the ESC to make an informed Go/ No-Go Launch Decision, walk them through the Launch Readiness summary (example provided later in this chapter) and the project team's launch recommendation. This will allow the ESC to look at specific data points about the different launch elements and understand any associated risk with launching.

> ## *Key Point*
>
> *This is the point in the project that you need to bring all the knowledge you've gained at a tactical level and communicate it at a strategic level to adequately capture the risks and opportunities with launching.*

Not only will this launch readiness assessment point out the next phase of work, it will help the team focus on the most important items first. At this point, your ESC is prepared to make a solid launch decision.

Remember how you always listed the team charter on your weekly project team meeting decks, and on your monthly ESC meeting decks? I recommend this because it's important to continually reiterate your main goal as you go through your project. I've been on some projects that by the time you reach the launch recommendation phase, you forget what your original goal was and have to cross your fingers and work backwards to make sure that what you provided at the end of the project meets the original goal.

Not only should we be cognizant of the fact that we are launching a new software product, we have to identify and manage the changes that will provide $1M in revenue within the first year, and $5M in revenue within three years. I continually refer back to the overall goal to make sure all the recommended changes outlined in the project plan lead us to the financial goals of the project.

In completing this assessment, you also need to lay out the immediate next steps required once the product is officially launched. I've seen too many launches in which executives vote to go forward with the launch, we then launch, and people walk away from the project thinking it's finished.

The following depicts a method I have used to guide a Go/No-Go Launch Decision.

- Create 'Launch Criteria' based upon what was identified earlier as required to launch; pick out only the critical items (month one)

- Follow the project plan leading up to testing (months one through four)

- Complete the testing of launch-critical items (months four through five)

- Work with Project Team to complete a detailed assessment of each Launch Criteria, documenting what is OK and what is Not OK at that point in time, and assessing the degree of risk in launching (months five through six)

- Work with Project Team to review the above assessment and decide on a Go / No-Go Recommendation (month six)

- Present Project Team's assessment and launch recommendation to the ESC to obtain a final launch decision (month six)

Here is an example of a Launch Readiness Assessment and Go / No-Go Launch Recommendation:

Go / No-Go Launch Recommendation

Prepared by the Project Team for the Executive Steering Committee

Launch Criteria – Customer-Facing:	Risk:
1. New software product is available in a pleasing, eye-catching package that reflects the product positioning	Low
2. Distribution channels are trained and new software is on the shelves	Low
3. New software is available on our website to view, obtain information, and buy (reflects product positioning)	Low
4. Customer is able to easily locate our new software within affiliate stores and on their websites	Low
5. Customer is able to easily purchase our new software	Low
6. Website homepage highlights the new software and pulls in new, qualified leads for our distribution teams	Low
7. Customer service is staffed and ready to take calls, chats, and emails	Low
8. Customer able to receive software within 1 business day, once ordered via our website	Low
9. Customer is able to easily return software, if necessary	Low

Launch Criteria – Back-Office:	Risk:
1. We are able to engineer new software with existing technology and manpower	Low
2. We are able to take orders from distributors	Low
3. We are able to distribute to appropriate distribution channels for sale to end users	Low
4. We are able to track orders, sales, returns, defects, customer service calls through existing financial systems	Low
5. We are able to provide CEO and senior management team with summary reports each month in same format as other product tracking reports	Low

Final Recommendation: Given the detailed assessment by the project team, leading to the summarized view above, **the Project Team recommends that we launch** on time, as planned.

Note: if the decision is made to launch, make sure to follow your project plan to ensure that the proper people are notified and the launch comes off without a hitch (move system changes into production, implement organizational changes, make process changes effective, notify employees, notify distributors and partners, launch advertising, etc.).

5 Secrets to *Guide* a

Faster Product Launch:

1. **G***round* Your Project with A Clear Goal

2. **U***nderstand* the Required Changes

3. **I***ntroduce* Tight Project Management

4. **D***rive* a Data-Based Launch Decision

5. **E***stablish* Pre- and Post-Launch Measurements

Chapter VII

The Fifth Secret:

Establish Pre- and Post-Launch Measurements

So far, we've covered four Secrets:

- First Secret: Ground Your Project with A Clear Goal

- Second Secret: Understand the Required Changes (Business Requirements)

- Third Secret: Introduce Tight Project Management

- Fourth Secret: Drive a Data-Based Launch Decision

Now, we're ready to cover the next step:

- Fifth Secret: Establish Pre- and Post-Launch Measurements

While preparing for launch, the first step was to articulate your goal. I mentioned earlier that you should sit down with your Finance team and translate your first year's annual goal into monthly views so you can begin to measure pre-post launch metrics.

Make sure you set up, prior to launch, the measurement vehicles which will allow you to collect data on key performance metrics on a monthly basis, otherwise this data will not be collected and you will loose credibility as a product launch leader. I've provided an example in Appendix E, Typical Post Launch Measurements.

Yes, this is another area that usually isn't the most appealing work, but you need to be able to provide key information to your ESC immediately upon launch. Don't be caught with not having the measurement vehicles to collect this critical information.

Key Point

Too often, companies spend too much

time working on their business models

to determine the amount of revenue

a new product should generate. While

I agree that financial forecasts are

important, keep in mind that they

are just that, forecasts, and that these

forecasts are only as good as the

assumptions on which they are based.

So, create monthly financial forecasts,

but keep in mind that the margin of

error can be pretty high if you have a

low confidence level in the underlying

assumptions that were used.

Along with setting up vehicles to measure what is stated in your project goal, you may also want to include softer data such as informal feedback from the following groups:

- Customers
- Employees involved in the process
- Partners or vendors involved in the process
- Distribution channels
- Steering committee

Remember: What gets measured gets managed, and what's managed happens, so you'll want to direct these efforts in the proper direction.

The Wrap Up

Chapter VIII

Closing Remarks

I hope you now understand the 5 Secrets to successfully manage and *guide* the launch of new products and that you are better prepared to lead your company's next product launch. These 5 Secrets provide a standardized approach to project managing new product launches that will help you manage both the work and the politics. They are based on Best Practices and years of learning what doesn't work.

Remember, what's properly managed is what actually happens.

Keep in mind that this Blueprint for Change is Not Just for New Product Launches...

These 5 Secrets can be applied in the business world to guide a business change as well as in your personal life to help guide a change in your life.

You can either follow the path that happens to pop up in front of you or you can create the path based upon where you want to go and tightly manage and guide your way to achieve your goals.

Thanks for your time. I hope you found this information informative and valuable.

If you have any comments or questions, please contact me at **Eliz@CavanaghConsulting.com** or visit my website at http://www.ElizabethCavanagh.com.

APPENDIX A

The 5 Secrets

Appendix A

5 Secrets to *Guide* a
Faster Product Launch:

1. **G***round* Your Project with A Clear Goal

2. **U***nderstand* the Required Changes

3. **I***ntroduce* Tight Project Management

4. **D***rive* a Data-Based Launch Decision

5. **E***stablish* Pre- and Post-Launch Measurements

APPENDIX B

Components of a Detailed Project Plan

Appendix B
Components of a Detailed Project Plan

Here's a list of areas that I look through when starting to create a detailed project plan. This is a comprehensive list, so everything might not apply to your project.

Possible components include:

1. Creating the Foundation for Change
   a. Strategic Foundation
      i. Write up business plan
      ii. Determine 'fit' with other product lines
      iii. Socialize with executives
      iv. Executives vote on funding
      v. Project sponsor/champion selected
      vi. Project team lead selected
   b. Set up Governance Structure
      i. Executive Steering Committee formed
         1. Identify ESC members
         2. Document goal of project
         3. Socialize project goal with ESC
      ii. Project Team formed
         1. Select team members
         2. Socialize project goal
         3. Document launch criteria
         4. Validate launch criteria with ESC

2. Identify and Track Key Dependencies
   a. Key Dependency #1
   b. Key Dependency #2
   c. Key Dependency #3...

3. Identify Changes Needed to Reach Project Goal
    a. New Product Description
        i. Detail out product description
        ii. Pricing
        iii. Promotion
            1. Promotion at Launch
            2. Promotion After Launch
        iv. Distribution
    b. Identify the Large Changes Needed to Reach Goal
    c. Identify Process Changes
        i. Customer-facing
        ii. Back-office facing
    d. Identify Organizational Changes
        i. Customer-facing
        ii. Back-office facing
    e. Identify System (Technology) Changes
        i. Customer-facing
        ii. Back-office facing
    f. Identify Long-Lead-Time Items
    g. Identify Interdependencies
    h. Complete Murphy's Law exercise
    i. Document the above information in Business Requirements Document

4. Create Project Management Tools to Manage Changes
    a. Team charter
    b. Launch criteria
    c. Project Plan
    d. High-level one-pager
    e. Detailed plan
    f. Issue Log & Action Item Log
    g. Weekly Project Team Meeting Deck
    h. Monthly Executive Steering Committee Meeting Deck

5. Make the Required PROCESS Changes
   a. Customer-facing Process Changes
      i. Advertising
         1. Prepare creative brief
         2. Create packaging for new product
         3. Identify launch offer
         4. Wrap a promotion around it
            a. TV
            b. Radio
            c. Print
            d. Web
            e. Pod-casts
            f. PR
            g. Internet search
            h. Social Media
            i. Direct mail
      ii. Distribution
      iii. Pricing
      iv. Packaging
      v. Purchasing
      vi. Returns
      vii. Referrals
      viii. Customer service
         1. Via phone – live person and/or chat
         2. Via phone – automated
         3. Via web – automated
   b. Back-Office Process Changes
      i. Account receivables
      ii. Account payables
      iii. Retailer pricing
      iv. Retailer contracts
      v. Budgets
      vi. Manufacturing
      vii. Quality control
      viii. Financial tracking

6. Make the Required ORGANIZATIONAL Changes
    a. Customer-facing Organizational Changes
        i. Assess number of new c/s reps to be hired
        ii. Write up job descriptions
        iii. Hire
        iv. Train
    b. Back-office facing Organizational Changes
        i. Assess number of new back-office people to be hired
        ii. Write up job descriptions
        iii. Hire
        iv. Train

7. Make the Required SYSTEM Changes
    a. Write business requirements
    b. Write technical requirements
    c. Write code
    d. Add in new requirements
    e. Test code
    f. Release code
    g. Note: Do all of the steps under 7) for each of these areas:
        i. Customer-facing
        ii. Advertising: website changes
        iii. Packaging development
        iv. Point of sale system changes
        v. Referral tracking system changes
        vi. Back-Office facing
        vii. Update invoicing tables
        viii. Update account receivables system
        ix. Update account payables system
        x. Update pricing database
        xi. Update retailer commission thresholds
        xii. Budget tracking
        xiii. Update GL tracking systems
        xiv. Update manufacturing machines
        xv. Quality control

8. User Testing
    a. Write user test cases
    b. Document pass/fail parameters
    c. Test
    d. Document test results
    e. Review test results and prioritize issues
    f. Fix top-priority issues
    g. Re-test

9. Prepare Launch Readiness Checklist
    a. List all 'required changes'
    b. Identify % complete for each area
    c. Identify risk with launching for each area (High, Med, Low)
    d. Identify person responsible for each area

10. Complete Training
    a. Create training materials
    b. Rollout training

11. Conduct Go / No-Go Launch Decision
    a. Project team reviews Launch Readiness Checklist
    b. Project team makes launch recommendation
    c. Exec Steering Committee makes final launch recommendation

12. Launch (assuming ESC decided to launch)
    a. Move system changes into production
    b. Make organizational changes effective
    c. Make process changes effective
    d. Notify employees
    e. Notify distributors
    f. Launch advertising

13. Post-Launch Measurement
   a. Identify success metrics
   b. Identify measurement vehicles
   c. Put measurement vehicles in place PRIOR to launch
   d. Conduct post-launch measurement
      i. After 1st day of launch
      ii. After 1st week of launch
      iii. After 1st month of launch
      iv. Every month
   e. Make changes based on pre-post measurements

APPENDIX C

Weekly Project Team Meeting Deck Example

Appendix C:

Weekly Project Team Meeting Deck Example

On the next page is an example of what I've used to guide our weekly project team meetings. The intent is to keep your team focused on what you want to be discussed without getting derailed.

It's been my experience that by providing a document that is visually appealing and high-level (vs. using a detailed project plan), people are more apt to attend your meetings and actively participate.

Software for Business Inc.

'The Tracker' Project Team Weekly Meeting

June 9, 2012

Project Charter

Ensure that Software for Business Inc launches a new software package called 'The Tracker' on October 15, 2012 (six months from now) in order to capture the untapped market of small and medium businesses that are not able to easily track their expenses.

By leading the launch of this product, we will enable our company to capture $1M in incremental revenue within a year, and $5M within 3 years.

Software for Business Inc.

High-Level Schedule

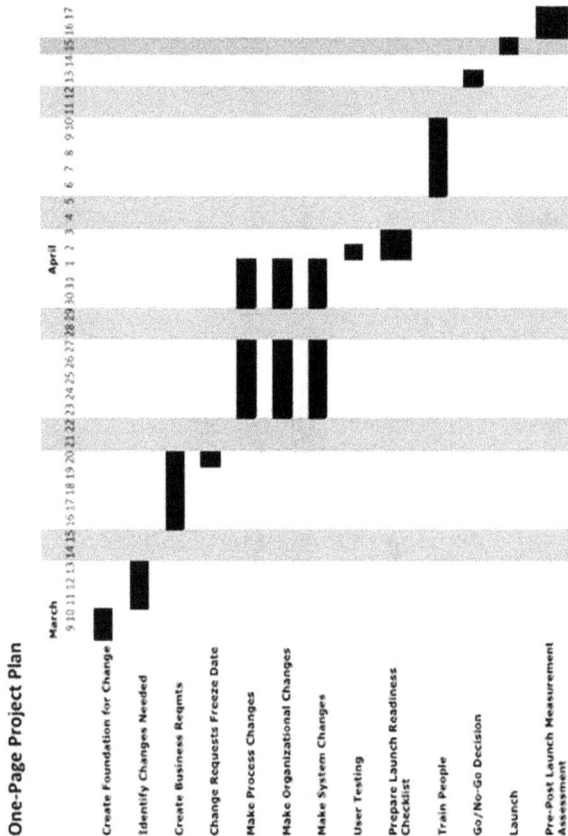

One-Page Project Plan

	March				April			
	9 10 11 12 13 14 15	16 17 18 19 20 21 22	23 24 25 26 27 28 29	30 31 1 2	3 4 5	6 7 8	9 10 11 12 13 14 15 16 17	

Create Foundation for Change

Identify Changes Needed

Create Business Reqmts

Change Requests Freeze Date

Make Process Changes

Make Organizational Changes

Make System Changes

User Testing

Prepare Launch Readiness Checklist

Train People

Go/No-Go Decision

Launch

Pre-Post Launch Measurement Assessment

Software for Business Inc.

What's Changed Since We Last Met

- Added a new requirement to do view reports from iPhones
- Solved the outstanding issue with testing environment access
- Launch date moved out 2 days

Software for Business Inc.

5

Project Status

Area:	Status:
Scoping	Green
Business Requiements	Yellow
Marketing Prep	Green
System Development	Yellow
Integration	Green
Training	Green
Process Changes	Green
Staffing Changes	Green
Testing	Yellow
Go / No-Go Decision	Green
Post-Launch Measurement	Green

Software for Business Inc.

Key Issues

Issue	Lead	Priority	Status	Updates
Engineering: Unable to secure the engineering resources required	Henry	High	Yellow	5/12/12: Henry spoke to his boss who will re-prioritize work in order to make resources available within 3 days 5/10/12: New issue
Marketing: New packaging vendor is unable to work on prototype of packaging to meet our dates	Nicole	Medium	Green	5/10/12: Nicole speaking with our current vendor who most likely will be able to make our dates if we move quickly 5/8/12: New issue

Software for Business Inc.

7

Completed & Upcoming Milestones

Completed Milestones:
- Additional funding for advertising
- Financial tracking in place

Upcoming Milestones:
- System testing begins on 6/12/12
- Creative brief complete on 6/14/12

Software for Business Inc.

8

Top Action Items

Action Item	Project/Lead	Priority	Status	Updates
Obtain new photos for website content from our ad agency	Elizabeth	Medium	Green	4/28/12 Eliz to provide by 5/20/12

Software for Business Inc.

9

Appendix

Software for Business Inc.

10

Scope of Project

What's In Scope:

Customer Facing: Target audience awareness, customer purchase, customer use, customer support, advertising, customer feedback

Back-Office Facing: engineering, testing, distribution, pricing, product management, marketing management, business planning, budgets, legal contracts, finance (billing and receivables)i, customer service

What's Out of Scope:

Customer Facing: other product advertising

Back-Office Facing: other product lines

Software for Business Inc.

Key
Assumptions/Constraints/Risks/
Dependencies

- Team will get proper staffing & funding
- IT will be able to meet software development changes on schedule
- Vendors will be able to supply needed components
- Product will be distributed through existing distribution channels
- Product will be priced at a mid-market price
- Product will be marketed as all new products are marketed, though targeted only to small and medium businesses.
- Competitor will be launching their product next year
- Target audience has a strong need for this type of software product

Software for Business Inc.

12

Appendix D

Summary of Best Practice Techniques and Tools

Appendix D

Summary of Best Practice Techniques and Tools

- **Seven Best Practice Techniques (to Manage your Project Team)**
  1. Cross-functional Work
  2. Full Disclosure of Information
  3. Accountability
  4. Tight Project Management of the Changes
  5. Weekly Project Team & Monthly Executive Steering Committee Meetings
  6. Supportive Leadership
  7. Constant Prioritization

- **Seven Best Practice Tools (to Manage the Work)**
  1. Team Charter
  2. Launch Criteria
  3. Project Plans
  4. Issues & Action Item Log
  5. Jeopardy Notice
  6. Succinct Weekly Project Team Meeting Deck
  7. Succinct Monthly Executive Steering Committee Deck

Appendix E

Typical Post Launch Measurements

Typical Post-Launch Tracking Elements

Example: Launching a New Software Product

	Month 1 June	Month 2 July	Month 3 Aug	Month 4 Sept	Month 5 Oct	Month 6 Nov	Month 7 Dec	YTD 2012
# Units Sold Data:								
# Software Packages Sold								
# Returns								
Net # Packages Sold								
Cumulative Net Packages Sold								
Revenue Data:								
Average Rev Per Package								
Total Revenue								
Average Commission/Package								
$ Value of Returns								
Net Revenue								
Operations Data:								
# Customer Service Calls								
Length of Call								
Calls/Package Sold								
Cost/Call								
Distribution Channel Mix:								
Direct Channel Sales								
Indirect Channel Sales								
% Direct Sales								
% Indirect Sales								

Appendix F

List of Figures

Appendix F

List of Figures

Figure	Page #

Appendix G

Summary of Key Points

Appendix G

Summary of Key Points

Key Point	Page #
Expect Issues; Manage Accordingly	33
Clarify Team Roles	40
Identify Long-Lead Time Tasks	53
Identify Changes Needed Early On	58
Manage Cross-Functionally	72
Identify Issues Early On	74
Create Team Ground Rules w/Team	76
Create Hi-Level & Detailed Project Plan	79
Tightly Manage Weekly Team Meetings	81
Have Solid Product Positioning	91
Weekly One-on-Ones w/Key Leads	97
Categorize ALL Project Information	103
Utilize Your Exec Steering Committee	109
Capture All Risks/Opportunities for Launch Recommendation	117
Create Monthly Forecasts with Ramp-Up	127

About the Author

Elizabeth Cavanagh is a Marketing executive with over 20 years experience in creating and launching high-tech products and marketing programs.

She has worked with a wide range of technology companies:

- Fortune 500 firms including Verizon Wireless, AirTouch Cellular and Pacific Telesis International
- Progressive start-ups including Virgin Mobile USA, WebEx Communications, ClipSync, and Globalstar Satellite Services.

She holds an MBA from Pepperdine University and a BSBA in Marketing from the University of Arizona.

Although Elizabeth has a wide variety of high-tech marketing experience (marketing strategy, business planning, product management, customer retention, market research, and IT), she added the most value for her clients when she helped them quickly launch new products.

With this in mind, ten years ago, Elizabeth started Cavanagh Consulting to help clients organize their teams around product launches. When clients ask what value she can provide, she replies, **"I bring structure to chaos, which allows teams to be much more productive, and allows companies to launch new products faster."** She has found that many companies, large and small, established and in start-up mode, have this same need.

Elizabeth has gathered a collection of **Best Practices** which allow teams to be more productive and to avoid the common mistakes most project managers make when launching new products. She wanted to pass on these best practices to others, which is how this book, *How to Launch a New Product Faster: 5 Secrets to Tight Project Management, Comprehensive Project Plans and Effective Change Management,* came about.

Elizabeth resides in the San Francisco Bay Area with her husband and their two children. You can find more information about Elizabeth in her website: **http://www.ElizabethCavanagh.com**.

ELIZABETH CAVANAGH
MARKETING & PRODUCT LAUNCH CONSULTANT

Bibliography

Allen, David. <u>Getting Things Done: the Art of Stress-Free Productivity</u>. London: Penguin Books Ltd. 2001.

Bossidy, Larry and Charan, Ram. <u>Execution: The Discipline of Getting Things Done.</u> New York: Crown Business. 2002.

Coyne, John. <u>Opposite Marketing: Why your Marketing is Backwards and How to Fix It!</u> Phoenix: Pivot Business Resources. 2009.

Isaacson, Walter. <u>Steve Jobs</u>. New York: Simon & Schuster. 2011.

Strother-Highfill, Donna. <u>Real People, Real Change: Stories of a Change Warrior in the Business World</u>. Richmond: Piping Tree Publishing. 2011.

Notes

Notes

Notes

Notes

www.ingramcontent.com/pod-product-compliance
Lightning Source LLC
Chambersburg PA
CBHW060526210326